TWO ANOMALOUS MANIFESTATIONS OF GRAVITY

Albert W. McKinney III
2019 September 6

DEDICATION

To Phyllis McKinney, my wife of 69 years.

Copyright © 2019 by Albert W. McKinney III

Library of Congress Cataloging-in-Publication Data
McKinney, Albert William III (1929-)
    Two anomalous manifestations of gravity.
    Kindle Direct Publishing, North Charleston, South Carolina
    ISBN: 9781689012324

TABLE OF CONTENTS

PAGE	ITEM	CONTENT
v	Preface	
1	How particle communications occur	
1	What happens in one cycle?	
2	What is gravity?	
3	The noncentric equation	
5	What is a gravitational offset?	
6	What causes a gravitational offset?	
7	Determining a gravitational offset	
7	The Sun	
11	The binary pulsar B1913+16	
15	Appendix	
15		Differential equations of motion
17		Vis viva integral
19		The main differential equation
20		Approximation 1
21		Approximation 2
25	References	

PREFACE

In 1687, Newton published his theory of gravity.

In 1859, Urbain Le Verrier used his exceedingly accurate observations to find that Newton's law seemingly did not explain the observed precession of the perihelion of Mercury's orbit. He decided that there must be a flaw in Newton's formula.

As will be shown later in this report, Le Verrier was not correct; Newton's formula is extremely accurate. The problem with Le Verrier's work is that he made an incorrect assumption, one which Newton himself had made, about the orbit of Mercury.

It is true that Mercury's orbit is essentially an ellipse. What Newton and Le Verrier assumed was that one focus of that ellipse was at the center of mass of the Sun. This assumption is very often used when analyzing the orbits of the planets. Hence any other assumption will be called anomalous.

In 2011 or so, it was discovered that the correct situation is that the operation of gravity in the solar system is anomalous. The purpose of this paper is to show how it is anomalous, and to give another example in which gravity also behaves anomalously.

This book deals with the particles in the universe. There are four kinds of particles: electrons, protons, positrons, and antiprotons.

Particles communicate with one another. The only form of communication to be considered here is between two particles. A simple form of communication occurs when one particle hits another. But most communications between particles occur between two particles which are not close to each other. Such communications are either electromagnetic or gravitational.

HOW PARTICLE COMMUNICATIONS OCCUR

How can two particles which may be inches or light years apart manage to make contact with each other? Of all the conceivable mechanisms to explain this, the simplest makes use of the following fact:

All particles vibrate; that is, they have frequencies. (It is known that electrons and positrons at rest have a frequency of 1.2×10^{20} cycles per second, while protons and antiprotons at rest have a frequency of 2.3×10^{23} cycles per second. By special relativity, moving particles have higher frequencies.)

WHAT HAPPENS IN ONE CYCLE?

Since particles vibrate, it is clear that they are not fixed lumps, but instead are composed of extremely malleable matter. At the beginning of a cycle, a particle has its full mass. (Henceforth, "mass" will be used to denote the mass and energy combined.) But the bulk of that mass is traveling at 299,792,458 meters per second (*light speed*).

Most of that mass (called a *probe*) is ejected in some random direction. As the probe passes through space, it leaves behind a minute trace of mass. In general, it travels in that direction until it runs out of mass, at which point it reverses direction and returns to its particle, gathering up the mass it had left behind. At this time, the particle again possesses its full mass. This ends the cycle. (Thus a particle behaves sort of like a yoyo.) The next cycle begins immediately, with the probe going out in almost the opposite direction.

However, it can happen that the probe hits another particle on its outward journey. In that case, there is an electromagnetic interaction between the probe and the particle (an exchange of an impulse of mass between the probe and the hit particle). This impulse is either attractive or repulsive, depending on the electromagnetic signs of the two particles involved. At the conclusion of the exchange, the probe ends its outward journey and returns to its particle, thus ending the current cycle.

It can also happen that the probe hits another probe on its outward journey. In that case, there is a gravitational interaction between the two probes (an exchange of an impulse of mass between the two probes). This impulse is *always* attractive. At the conclusion of the exchange, both probes continue on their outward journeys, but in slightly different directions.

WHAT IS GRAVITY?

Gravity between two objects is simply the accumulation of the forces of gravitational interactions between pairs of particles, one from each object. (A)

Newton represented this very neatly in his classical formula

$$F = \frac{GM_1M_2}{r^2},$$

where F is the gravitational attraction between two bodies of masses M_1 and M_2, G is Newton's gravitational constant, and r is the distance between the centers of mass of the two bodies. This is equivalent to statement (A) above. To see this, simply replace the masses of the two bodies by the sum of all the particles in those masses:

$$F = \frac{G(m_{11} + m_{12} + ... + m_{1M})(m_{21} + m_{22} + ... + m_{2N})}{r^2}.$$

The product of the two sums includes one term for each pair of particles with one from each mass.

Thus it is clear that for smaller bodies (such as planets), Newton's formula is *exact*.

THE NONCENTRIC EQUATION

The problem posed by Le Verrier's data was not due to an error in Newton's law, but instead was due to an incorrect use of that law. Le Verrier had assumed (as had Newton) that every planet had an elliptical orbit in which one focus was at the center of mass of the Sun.

Several years ago (about 2011), I began to suspect that that assumption was wrong. I derived an equation (the *noncentric equation*) which deals with the paths of two orbiting objects (such as the Sun and a planet, or two stars orbiting each other). Each of these two orbits is (very close to) an ellipse with two focal points, one of which is in the other object. Say these points are P_1 and P_2. Classically these points are assumed to be the centers of mass of these objects. That is not assumed here. The only assumptions are that P_1 and P_2 lie on a straight line from the center of mass of object 1 to the center of mass of object 2; and that the distance between P_1 and P_2 equals the distance between the centers of mass of the two objects minus the sum of the gravitational offsets of the two objects.

The equation was derived using conventional celestial mechanics in Euclidean space. The equation and its derivation are found in the Appendix.

One form of the equation relates the shift s of the precession of the periastron on one orbit of the planet to the gravitational offset q of the star:

(1) $$s = 360 \times 3600 \left(\frac{1}{\sqrt{1-2Lq}} - 1 \right)$$

in arcseconds per orbit.

A second form gives the equation of motion of one of the objects:

(2) $$p = \frac{1}{A\cos\left(\sqrt{1-2Lq}\,\theta - \omega\right) + \frac{L}{1-2Lq}}$$

In these equations,

A and ω are constants specific to the object;
p is the distance of the object from the apparent center of mass of the star when the object is at a point θ in its orbit;
q is the value of the gravitational offset of the star;

and

$$L = \frac{G(M_1 + M_2)T^2}{4\pi^2 a^4 (1-e^2)},$$

where

G is the gravitational constant $(m^3 kg^{-1} s^{-2})$;
M_1 is the measured mass of the star (kg);
M_2 is the mass of the planet or the other star (kg);
T is the time of one orbit (s);
a is the semimajor axis of the ellipse (m);
e is the eccentricity of the ellipse.

WHAT IS A GRAVITATIONAL OFFSET?

Many (perhaps all) stars have a gravitational offset. This is a situation in which (1) a star A has a planet P (or another star B) in orbit about it, and (2) the planet P (or the other star B) has an elliptical orbit in which one focus is close to the center of mass of star A, but is not at that center of mass (it is at the *apparent* center of mass of star A). The following sketch (which is not to scale) shows an example. The gravitational offset in the sketch is the distance between the true center of mass and the apparent center of mass of the star, as perceived by the planet.

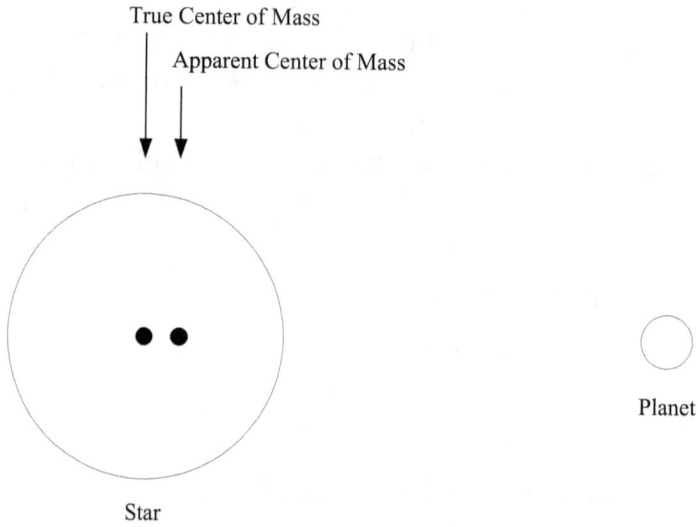

This sketch applies to the case where the second object is a planet P. Another case is where a second star B appears in place of the planet P. In either case, the apparent center of mass refers only to what the planet P or the second star B perceives, but not to any other object orbiting star A.

WHAT CAUSES A GRAVITATIONAL OFFSET?

A gravitational hit occurs when a probe hits another probe. Since probes travel at light speed, a probe generally spends very little time in an object. However, once in motion, a probe does not distinguish between targets. It connects with any other particle or probe that it encounters, whether it is in the same object or another object. Due to its speed, an encounter is usually in another object.

It is easy to see that the gravitational attraction between two objects is proportional to the number of gravitational hits between them. Newton's formula presumes that over each small period of time, there is a gravitational hit between every particle of object 1 and object 2.

But if a probe starts on one side of a star and heads toward the other side, it finds so many possible particles to encounter that it often hits another particle in the star, and thus never leaves the star to set up a gravitational encounter in another object.

The result is that many probes on one side of the star never emerge from the opposite side of the star. For this reason, a planet or companion star on that other side receives fewer than expected probes from the star, which causes the planet or companion star to have an elliptical orbit with a focus on a point offset from the true center of mass of the star.

DETERMINING A GRAVITATIONAL OFFSET

The noncentric equation can sometimes be used to calculate the value of the gravitational offset of a star. The equation shows the relation between the motions of either a star and its planet or two orbiting stars in the case where either or both objects have gravitational offsets.

THE SUN

Applied to the Sun, the equation gives the unique relation between the gravitational offset q of the Sun and the size s of the shift in the perihelion of a planet from one orbit to the next:

There are 8 planets and one asteroid for which the above values are known. Perhaps the most accurately known are those for Mercury:

$$G = 6.67384 \times 10^{-11} \ (m^3 kg^{-1} s^{-2})$$
$$M_1 = 1.9884 \times 10^{30} \ (kg)$$
$$M_2 = 3.3011 \times 10^{23} \ (kg)$$
$$T = 7,600,530.24 \ (s)$$
$$a = 57,909,100,000 \ (m)$$
$$e = 0.205\,630$$

These values yield $L = 1.80295 \times 10^{-11}$.

To get the value of p, I used the values from the Wikipedia article on the perihelion precession of Mercury (extracted on 2019 August 14): The observed precession is 574.10 ± 0.65 and the gravitational tugs of other solar bodies amounts to 532.3035, both measurements in arcseconds per Julian century. Leaving out the values for general relativity and two very small other values, the difference between the first two values gives the observed precession values due to things other than the other solar bodies. This difference is 41.7965 arcseconds per Julian century. To convert this value to arcseconds per orbit of Mercury, multiply by the seconds per orbit of Mercury, 7,600,530.24, and divide by the seconds per Julian century, 3,155,814,977. The result is $p = 0.100\,663\,557$.

Solving equation (1) for q yields

(3) $$q = \frac{1}{2L}\left[1 - \left(\frac{360 \times 3600}{p + 360 \times 3600}\right)^2\right]$$

Using the above values for L and p results in the value $q = 4{,}308$ meters.

The implication is that the Sun has a gravitational offset of 4,308 meters. That is, no solar planet has an orbit with a focal point at the center of mass of the Sun; rather, each such planet's orbit has a focal point which is 4,308 meters from the center of mass of the Sun. Of course, the offset is always applied along the straight line between the center of mass of the Sun and the center of mass of the planet.

I estimate that the figure 4,308 is accurate to 2 significant figures, since a previous calculation (2013?) from a corresponding Wikipedia article yielded the figure 4,375 meters.

These values of L and q lead to the value $\sqrt{1 - 2Lq} = 0.999\,999\,922$. With these values, equation (2) becomes

(4) $$p = \frac{1}{A\cos(0.999\,999\,922\theta - \omega) + 1.802\,949\,608 \times 10^{-11}}$$

At perihelion, the value of p is a minimum, and so the denominator is at a maximum. Since the value inside the parentheses increases with θ, the argument of the cosine is positive for higher values of θ. The maximum occurs when the cosine equals 1, and that is for any value of θ such that

$\theta = \dfrac{2\pi n + \omega}{\sqrt{1-2Lq}}$, where n is a nonnegative integer. From one perihelion to the next, θ goes from $\dfrac{2\pi n + \omega}{\sqrt{1-2Lq}}$ to $\dfrac{2\pi(n+1) + \omega}{\sqrt{1-2Lq}}$, which is a change of

$$\dfrac{2\pi}{\sqrt{1-2Lq}} = \dfrac{2\pi}{0.999\,999\,922} = 6.283\,185\,795.$$

In one orbit, starting at a perihelion, θ goes from $\dfrac{2\pi n + \omega}{\sqrt{1-2Lq}}$ to $\dfrac{2\pi n + \omega}{\sqrt{1-2Lq}} + 2\pi$, a change of 2π. This takes $7{,}600{,}530.24$ seconds. Thus the change of $\dfrac{2\pi}{\sqrt{1-2Lq}}$ takes $\dfrac{7\,600\,530.24}{0.999\,999\,922} = 7\,600\,530.83$ seconds. That is, the time between two perihelia is 0.59 seconds more than the time for one orbit.

    The above pages present proof that the Sun's gravity is anomalous, since the Sun has a gravitational offset of about 4,308 meters. Hence every planet that orbits the Sun has an elliptical orbit, one focus of which is a point within the Sun, and which lies on a straight line from the center of mass of the Sun to the center of mass of the planet, and lies 4,308 meters from the center of mass of the Sun towards the planet.

    As for Einstein's explanation of Le Verrier's findings, it turns out that general relativity is itself an approximation, albeit a fairly accurate one. But it is not exact. The justification for this statement is that a more accurate model of the solar system gives slightly more accurate values for the movements of the planets, and these more accurate values differ from the values given by general relativity. See Reference 1 for details.

THE BINARY PULSAR B1913+16

The following measurements of this binary system are taken from an article by Weinberg and Taylor (see Reference 2). The system consists of two pulsars which orbit each other every 323 days or so.

Semimajor axis of larger star:

$$a_1 = \frac{2.341\,772\,5c}{0.734} = 956,466,000 \text{ meters}$$

Eccentricity $\quad e_1 = 0.617\,133\,8$

Period of orbit $\quad T_1 = 27,906,979.59$ seconds

Mass of larger star $\quad M_1 = 1.441\,4\,M_s$

Mass of smaller star $\quad M_2 = 1.386\,7\,M_s$

Shift in periastron $\quad \omega' = 4.226\,595$ degrees / year

where $c = 299,792,458$ m s^{-1} is the velocity of light in a vacuum, and $M_s = 1.988\,4 \times 10^{30}$ kg is the mass of the Sun.

The shift in periastron per orbit is equal to

$$\frac{4.226\,595 \times 27,906,979.59}{31,558,149.8} = 0.003\,737\,592\,39,$$

and this is equal to $360\left(\dfrac{1}{\sqrt{1-2Lq}}-1\right)$, where q is the sum of the gravitational offsets of the two pulsars. Thus,

$$\dfrac{1}{\sqrt{1-2Lq}}-1 = \dfrac{0.003\,737\,592\,39}{360} = 1.038\,220\,107 \times 10^{-5} \equiv \varepsilon,$$

so that

$$1-2Lq = \dfrac{1}{(1+\varepsilon)^2} = 0.999\,979\,236,$$

and thus

$$Lq = \dfrac{1}{2}\left(1-\dfrac{1}{(1+\varepsilon)^2}\right) = 1.038\,203\,939 \times 10^{-5}.$$

Formula (A10) in the Appendix implies the following:

$$L = \dfrac{1-2Lq}{a(1-e^2)} = \dfrac{0.999\,979\,236}{956,466,000 \times \left(1-(0.617\,133\,8)^2\right)}$$
$$= 1.688\,606\,526 \times 10^{-9}$$

Hence

$$q = \dfrac{1.038\,203\,939 \times 10^{-5}}{1.688\,606\,526 \times 10^{-9}} = 6{,}148 \text{ meters}.$$

Since the masses of the two pulsars are nearly the same, the individual gravitational offsets are roughly

$q_1 = 3,133$ meters and $q_2 = 3,105$ meters. Since both stars are neutron stars, their radii are both about 10 kilometers. Thus their gravitational offsets are about a third of their radii. The gravitational attraction between the two pulsars is hence due to the attraction between (very roughly) half the mass of each.

And this concludes the second example of the anomalous manifestation of gravity.

Based on these two examples, it seems reasonable to conclude that most stars have gravitational offsets.

APPENDIX

Proofs of Equations (1) and (2)

Differential Equations of Motion

The key idea which explains the shift in the perihelia of planets is this: When two bodies are in orbit around each other, two of the things which determine their orbits are not their centers of mass, but instead their apparent centers of mass. They sense each other's presence gravitationally, but each senses the center of mass of the other at a point possibly offset from its true center of mass. These offsets determine their mutual orbits.

Suppose two objects of gravitational masses M_1 and M_2 move in orbits around each other. Denote the positions of their centers of mass by the vectors $\mathbf{P}_1(t)$ and $\mathbf{P}_2(t)$. Let $\mathbf{P} = \mathbf{P}_2 - \mathbf{P}_1$, let $p = |\mathbf{P}|$, and let q_1 and q_2 be the sizes of the gravitational offsets of the two objects. Let $\mathbf{S}_1$ be the point between $\mathbf{P}_1(t)$ and $\mathbf{P}_2(t)$ which is a distance of q_1 from $\mathbf{P}_1(t)$, and let $\mathbf{S}_2$ be the point between $\mathbf{P}_1(t)$ and $\mathbf{P}_2(t)$ which is a distance of q_2 from $\mathbf{P}_2(t)$.

These assumptions are made:

1. Each of the two objects acts as a rigid body.
2. $\mathbf{P}_1$, $\mathbf{S}_1$, $\mathbf{S}_2$, and $\mathbf{P}_2$ lie on a straight line in that order.
3. The offset distances q_1 and q_2 do not change with time.

Let $q = q_1 + q_2$.

The gravitational attraction between the two objects is applied at the two points S_1 and S_2. By the above assumptions, the distance between them is equal to $p - q$. Let F_i denote the gravitational force on object i; then (by analogy with Newton's law)

(A1) $$\mathbf{F}_1 = \frac{GM_1M_2}{(p-q)^2} \frac{\mathbf{P}}{p}$$

and $\mathbf{F}_2 = -\mathbf{F}_1$.

By the first and second assumptions, the forces act on the gravitational masses, and the force vectors are applied at the apparent centers of mass. (One result is that the force on the apparent center of mass of the Sun contributes to the rotation of the Sun.) Hence, the differential equations of motion are

(A2) $$M_i \mathbf{P}_i'' = \mathbf{F}_i,$$

where primes denote time derivatives. Since $\mathbf{F}_2 = -\mathbf{F}_1$,

$$M_1 \mathbf{P}_1'' + M_2 \mathbf{P}_2'' = 0,$$

and so the center of mass of the entire system:

$$\frac{M_1 \mathbf{P}_1 + M_2 \mathbf{P}_2}{M_1 + M_2}$$

moves in a straight line.

Now consider the differential equation for P. Since $\mathbf{P} = \mathbf{P}_2 - \mathbf{P}_1$, it follows from equation (A2) that

(A3) $\quad \mathbf{P''} = \mathbf{P_2''} - \mathbf{P_1''} = \dfrac{\mathbf{F_2}}{M_2} - \dfrac{\mathbf{F_1}}{M_1} = \dfrac{-K\mathbf{P}}{p(p-q)^2},$

where

$$K = G(M_1 + M_2).$$

Taking the vector product of P with the first and last sides of equation (A3), it is found that $\mathbf{P} \times \mathbf{P''} = 0$, which can be integrated directly to yield the fact that $\mathbf{P} \times \mathbf{P'}$ equals a constant vector. Thus, the orbits lie in a plane perpendicular to that vector. Take that plane as the x-y plane, and let the origin be at $\mathbf{P_1}$.

Vis Viva Integral

Put equation (A3) in rectangular coordinates:

$$x'' = \dfrac{-Kx}{p(p-q)^2},$$

$$y'' = \dfrac{-Ky}{p(p-q)^2}.$$

Multiply the first equation by $2x'$, the second by $2y'$, and add:

(A4) $\quad 2x'x'' + 2y'y'' = \dfrac{-2K(xx' + yy')}{p(p-q)^2}.$

But by definition, the square of the velocity, v^2, is given by

$$v^2 = x'^2 + y'^2,$$

and so the left side of (A4) equals $(v^2)'$. Also, $p = \sqrt{x^2 + y^2}$ and q is a constant. Hence,

$$(p-q)' = p' = \frac{2xx' + 2yy'}{2\sqrt{x^2 + y^2}} = \frac{xx' + yy'}{p},$$

and so the right side of (A4) is equal to

$$\frac{-2K(p-q)'}{(p-q)^2}.$$

Thus, equation (A3) can be integrated to obtain

$$v^2 = \frac{2K}{p-q} + c_1$$

for some constant c_1.

The Main Differential Equation

Using polar coordinates, set

$$\mathbf{P} = p(\cos\theta, \sin\theta).$$

Then the second derivative is

$$\mathbf{P}'' = (p'' - p\theta'^2)(\cos\theta, \sin\theta) \\ + (2p'\theta' + p\theta'')(-\sin\theta, \cos\theta),$$

and the differential equation (A2) becomes

$$\left(p'' - p\theta'^2 + \frac{K}{(p-q)^2}\right)(\cos\theta, \sin\theta) \\ + (2p'\theta' + p\theta'')(-\sin\theta, \cos\theta) = 0.$$

Since $(\cos\theta, \sin\theta)$ and $(-\sin\theta, \cos\theta)$ are mutually perpendicular unit vectors, the above sum can vanish only if the coefficients of those vectors are each zero; thus

(A5) $$p'' - p\theta'^2 + \frac{K}{(p-q)^2} = 0$$

and

$$2p'\theta' + p\theta'' = 0.$$

From the latter equation, it follows that

$$(p^2\theta')' = p(2p'\theta' + p\theta'') = 0,$$

and so for some constant h,

(A6) $\quad p^2\theta' = h$.

Using this fact in equation (A5) yields

(A7) $\quad p'' - \dfrac{h^2}{p^3} + \dfrac{K}{(p-q)^2} = 0$.

Next, use the standard transformation $p = 1/u$, and let the angle θ replace time as the independent variable, with dots indicating derivatives with respect to θ. By the definition of u, along with equation (A6), it follows that

$$p' = \dfrac{-u'}{u^2} = \dfrac{-\dot{u}\theta'}{u^2} = -\dot{u}p^2\theta' = -h\dot{u}.$$

One more use of equation (A6) yields

$$p'' = -h\ddot{u}\theta' = -h^2 u^2 \ddot{u}.$$

Hence, rewriting equation (A7) in terms of u rather than p and dividing the result by $-h^2 u^2$ yields

(A8) $\quad \ddot{u} + u = \dfrac{L}{(1-qu)^2}$,

where

$$L = \dfrac{K}{h^2}.$$

Approximation 1

Equation (A6) represents twice the rate of accumulation of area in polar coordinates. The integral of

(A6) over one orbit (say for $t = 0$ to T, where T is the time required for the orbit) yields twice the area contained within the orbit. Under the (usually very accurate) approximation that the orbit is an ellipse, that area equals πab, where a and b are the lengths of the semimajor and semiminor axes. Hence, the integral results in the equation

$$2\pi ab = Th,$$

so that

$$h = \frac{2\pi ab}{T} = \frac{2\pi a^2 \sqrt{1-e^2}}{T},$$

where e is the eccentricity of the ellipse. Consequently,

$$L = \frac{K}{h^2} = \frac{G(M_1 + M_2)T^2}{4\pi^2 a^4 (1-e^2)}.$$

Approximation 2

Each offset q_i is a fraction of the radius of the corresponding star or planet, and so the sum of the offsets, q, is much smaller than the separation p between the two objects. Thus the quantity $q/p = qu$ is much smaller than 1, and so for all u of interest, the denominator on the right side of equation (A8) can be expanded into a power series which converges rapidly. In fact, an excellent approximation is obtained by dropping all terms of order u^2 and higher. The result is the approximate equation

$$\ddot{u} + u \cong L(1 + 2qu),$$

or

$$\ddot{u} + (1-2Lq)u \cong L.$$

The nature of the solution to this equation depends on whether $2Lq$ is smaller or larger than 1. It is not correct to say that $2Lq$ is small merely because qu is small. However, for the solution to be at all meaningful, $2Lq$ must be less than 1; otherwise, the solution would not be periodic, but would either expand or contract exponentially, which is not of interest. Hence for the moment, assume that $2Lq$ is less than 1; it is easy to show that it is positive. This leads to the solution

$$u = A\cos\left(\sqrt{1-2Lq}\,\theta - \omega\right) + \frac{L}{1-2Lq}$$

for constants A and ω, as is easily verified by differentiation. From this, it is possible to solve for p:

(A9) $$p = \frac{1}{A\cos\left(\sqrt{1-2Lq}\,\theta - \omega\right) + \dfrac{L}{1-2Lq}}.$$

For small values of $2Lq$, this equation is very nearly that of an ellipse. Thus equation (1), which is the same as equation (A9) above, is established.

Recall that p represents the distance of the true center mass of the second object from that of the first object in a coordinate system in which the origin is at the true center of mass of the first object. The minimum value of p occurs when the denominator in equation (A9) reaches its maximum, and that happens when the cosine equals 1.

Similarly, the maximum value of p occurs when the denominator reaches its minimum, and that happens when the cosine equals -1. Of course, this presumes that the denominator never vanishes for any value of θ. It will be seen below that this is the case.

For an ellipse, the separation p at the time of periastron is equal to $a(1-e)$, and at apastron, p equals $a(1+e)$. Using the fact that equation (A9) is very nearly that of an ellipse, form the equations

$$A + \frac{L}{1-2Lq} = \frac{1}{a(1-e)},$$

$$-A + \frac{L}{1-2Lq} = \frac{1}{a(1+e)}.$$

These can be combined to yield

(A10) $\quad \dfrac{L}{1-2Lq} = \dfrac{1}{a(1-e^2)}$

and

$$A = \frac{e}{a(1-e^2)}.$$

The interval between two periastra is found by taking two successive values of θ for which the argument of the cosine is equal to a multiple of 2π. Two such values are θ_1 and θ_2, where $\theta_1 < \theta_2$, and where

$$\sqrt{1-2Lq}\,\theta_1 - \omega = 0$$

and

$$\sqrt{1-2Lq}\,\theta_2 - \omega = 2\pi$$

for some value ω.

The change in θ from one periastron to the next is thus

$$\theta_2 - \theta_1 = \frac{2\pi + \omega}{\sqrt{1-2Lq}} - \frac{\omega}{\sqrt{1-2Lq}}$$

$$= \frac{2\pi}{\sqrt{1-2Lq}}.$$

If this change were equal to 2π, there would be no shift in periastron. Otherwise, the shift in periastron is equal to the above amount minus 2π:

$$2\pi\left(\frac{1}{\sqrt{1-2Lq}} - 1\right) \quad \text{in radians}$$

$$360\left(\frac{1}{\sqrt{1-2Lq}} - 1\right) \quad \text{in degrees}$$

(A11) $\quad 360\times 3600\left(\dfrac{1}{\sqrt{1-2Lq}} - 1\right) \quad$ in arcseconds.

Thus equation (2), which is the same as equation (A11) above, is established.

REFERENCES

1. Albert W. McKinney III, *A Better Model of the Solar System, and Why the Universe is Not Expanding*, CreateSpace Independent Publishing Platform, 2019 April 22.

2. J. M. Weisberg and J. H. Taylor, Relativistic Binary Pulsar B1913+16: Thirty Years of Observations and Analysis, http://arxiv.org/PS_cache/astro-ph/pdf/0407/0407149v1.pdf.

www.ingramcontent.com/pod-product-compliance
Lightning Source LLC
Chambersburg PA
CBHW070846220526
45466CB00002B/900